WACO-MCLENNAN C
1717 AUSTIN AVE
WACO, TX   76701

Animals in My Yard

Armadillos

by Amy McDonald

Blastoff! Beginners are developed by literacy experts and educators to meet the needs of early readers. These engaging informational texts support young children as they begin reading about their world. Through simple language and high frequency words paired with crisp, colorful photos, Blastoff! Beginners launch young readers into the universe of independent reading.

Sight Words in This Book

a	have	no	the	to
an	in	one	their	up
day	is	out	them	use
eat	it	she	these	
for	long	some	they	
go	make	that	this	

This edition first published in 2021 by Bellwether Media, Inc.

No part of this publication may be reproduced in whole or in part without written permission of the publisher. For information regarding permission, write to Bellwether Media, Inc., Attention: Permissions Department, 6012 Blue Circle Drive, Minnetonka, MN 55343.

Library of Congress Cataloging-in-Publication Data

Names: McDonald, Amy, author.
Title: Armadillos / by Amy McDonald.
Description: Minneapolis, MN : Bellwether Media, 2021. | Series: Blastoff! beginners : Animals in my yard | Includes bibliographical references and index. | Audience: Ages PreK-2 | Audience: Grades K-1
Identifiers: LCCN 2020029476 (print) | LCCN 2020029477 (ebook) | ISBN 9781644873595 (library binding) | ISBN 9781648340604 (ebook)
Subjects: LCSH: Armadillos--Juvenile literature.
Classification: LCC QL737.E23 M34 2021 (print) | LCC QL737.E23 (ebook) | DDC 599.3/12--dc23
LC record available at https://lccn.loc.gov/2020029476
LC ebook record available at https://lccn.loc.gov/2020029477

Text copyright © 2021 by Bellwether Media, Inc. BLASTOFF! BEGINNERS and associated logos are trademarks and/or registered trademarks of Bellwether Media, Inc.

Editor: Christina Leaf   Designer: Jeffrey Kollock

Printed in the United States of America, North Mankato, MN.

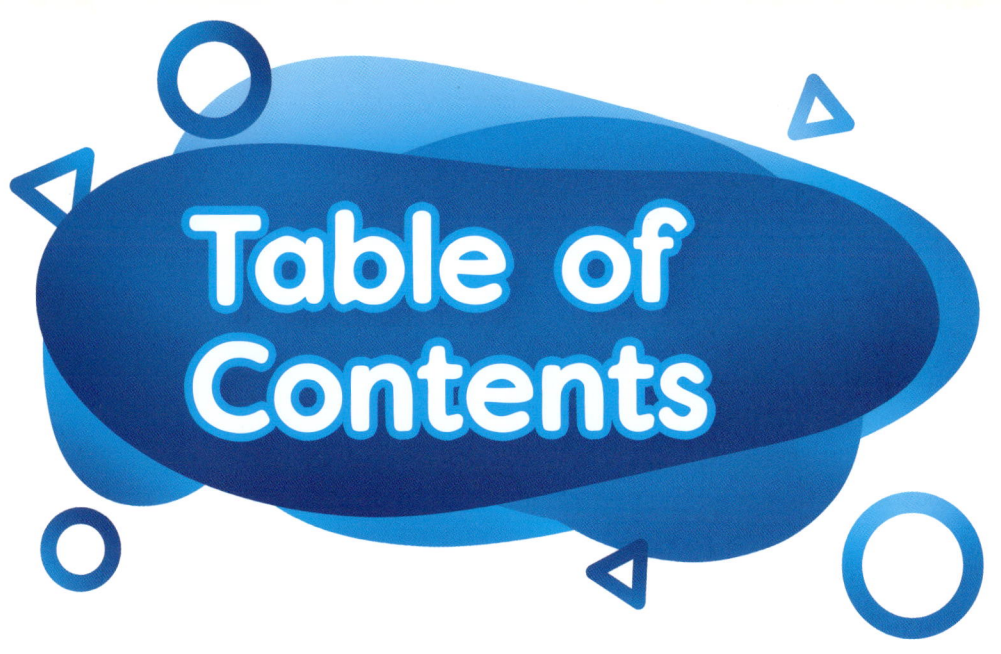

Table of Contents

Armadillos!	4
Body Parts	6
The Lives of Armadillos	12
Armadillos Facts	22
Glossary	23
To Learn More	24
Index	24

Armadillos!

Is that a ball? No!
It is an armadillo!

Body Parts

Armadillos have hard **plates**. These **protect** them.

plates

They have a
long tongue.
It snaps up food.

tongue

They have claws.
They dig
in the dirt.

claws

The Lives of Armadillos

Armadillos live in **burrows**. They dig to make homes.

burrow

Armadillos sleep a lot. They sleep 16 hours a day!

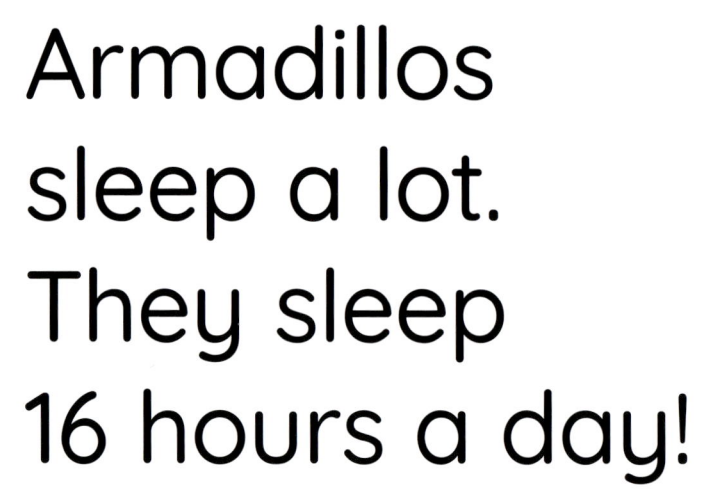

They go out for food.
They use their nose to hunt.

nose

17

Armadillos eat bugs. Some eat plants or eggs.

bugs

plants

eggs

This one digs for ants.
She licks them up.
Yum!

Armadillo Facts

Armadillo Body Parts

plates

nose

claws

tongue

Armadillo Food

bugs   plants   eggs

Glossary

burrows

the homes of armadillos

plates

hard coverings that keep armadillos safe

protect

to keep safe from harm

To Learn More

ON THE WEB

FACTSURFER

Factsurfer.com gives you a safe, fun way to find more information.

1. Go to www.factsurfer.com.
2. Enter "armadillos" into the search box and click 🔍.
3. Select your book cover to see a list of related content.

Index

ants, 20
bugs, 18
burrows, 12
claws, 10
day, 14
dig, 10, 12, 20
dirt, 10
eat, 18
eggs, 18, 19

food, 8, 16
homes, 12
hunt, 16
licks, 20
nose, 16, 17
plants, 18, 19
plates, 6
protect, 6
sleep, 14

tongue, 8, 9

The images in this book are reproduced through the courtesy of: lalito, front cover, p. 22; Eric Isselee, pp. 3, 4, 5, 6, 10, 23 (plates); Wildlife GmbH/ Alamy, pp. 6-7, 18-19; Danita Delimont/ Alamy, pp. 8-9; davemhuntphotography, pp. 10-11; Lindasj22, p. 12; Foto 4440, pp. 12-13; MyImages - Micha, pp. 14-15; Marcelo Morena, pp. 16-17; Meister Photos, p. 18; Burning Bright, p. 19 (plants); dcwcreations, p. 19 (eggs); Dreamframer, p. 20; Leena Robinson, pp. 20-21; corlaffra, p. 22 (bugs); Dark Caramel, p. 22 (plants); Jw5 Photography, p. 22 (eggs); Yashkin Ilya, p. 23 (burrows); belizar, p. 23 (protect).